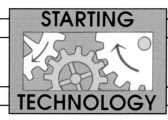

TIME

John Williams

Illustrated by
Malcolm S. Walker

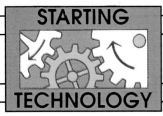

Titles in this series

AIR
COLOUR AND LIGHT
ELECTRICITY
FLIGHT
MACHINES
TIME
WATER
WHEELS

Words printed in **bold** appear in the glossary on page 30

© Copyright 1990 Wayland (Publishers) Ltd

First published in 1990 by
Wayland (Publishers) Ltd
61 Western Road, Hove
East Sussex BN3 1JD, England

Editor: Anna Girling
Designer: Kudos Design Services

British Library Cataloguing in Publication Data
Williams, John
  Time.
  1. Time
  I. Title II. Series
  529

ISBN 1 85210 922 X **Hardback**
ISBN 0 7502 0269 6 **Paperback**

Typeset by Kudos Editorial and Design Services, Sussex, England
Printed in Italy by Rotolito Lombarda S.p.A.
Bound in Belgium by Casterman S.A.

CONTENTS

Clocks	4
Day and Night	6
The Seasons	8
Sundials	10
More Sundials	12
Water Clocks	14
Old Water Clocks	16
Candle Clocks	18
Timers	20
Pendulums	22
Digital Clocks	24
Alarm Clocks	26
Notes for Parents and Teachers	28
Glossary	30
Books to Read	31
Index	32

STARTING TECHNOLOGY

CLOCKS

Clocks tell us the time. They tell us the time to get up in the morning and when to go to bed at night. They tell us when to go to school, or what time to catch a train or bus.

This clock has Roman numbers on it. What time is it saying?

Making a model clock

You will need:

A piece of white card
Scissors
Felt tip pens
A paper fastener
A cardboard box
Paper glue

1. Cut out a circle from the piece of card. You could draw round a plate to give you the right shape.

2. Draw in the numbers on the **dial**. Look at a real clock to see where they should go.

3. Cut two small strips of card for the hands of the clock. One should be shorter than the other. Fix them to the centre of the dial with the paper fastener.

4. Stick your dial on to the box. Decorate the front and sides of the box.

STARTING TECHNOLOGY

DAY AND NIGHT

Making a 'My Day' time line

You will need:

A piece of white card
Scissors
Felt tip pens
A ruler

1. Cut out a long piece of card. Divide it into sections using a pen and ruler.

2. Draw pictures in each section to show what you do during the day and night. Start with a picture of yourself asleep in bed.

Bed   Breakfast   School   Dinner   More school   Tea   Play

3. Fold the card along each line to make a zigzag book.

Making a 'My Day' picture clock

You will need:

Card or paper
Felt tip pens
A ruler

1. Draw a circle on a piece of card or paper.

2. In your circle draw little pictures to show what you do at different times during the day — just like you did with the time line.

3. Draw one for different days of the week and for the weekend. Are they the same?

Have you ever seen a clock like this one in a public park or garden? The dial is made from all kinds of flowers.

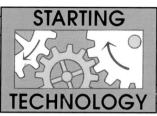

THE SEASONS

You can often tell what **season** it is just by looking out of your window. If you have a garden you can see what flowers are growing. Some, like daffodils, flower in the spring. Others, like poppies, flower in summer. If the leaves are falling off the trees, it may be that autumn has arrived. In winter, there may be frost and snow.

It is a bright, sunny day, but what things in this picture tell us it must be autumn?

Making a zigzag book of the seasons

You will need:

A piece of white card
Scissors
Felt tip pens
A ruler

1. Cut out a strip of card and divide it into sections to make a zigzag book. You can make one like this to show the four seasons of the year, or you can draw twelve sections — one for each month.

Winter   Spring   Summer   Autumn

2. Draw pictures in each section to show how the plants, animals and weather change during the year.

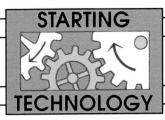

SUNDIALS

People have used the sun to keep time for thousands of years. The first sundials were made by the **Romans**.

Sundials are easy to make. Here are some different **designs**. You will need to check the shadows at regular times with a clock. In this way you can make a proper dial by marking the hours on it. You may have to wait for a sunny day!

WARNING: Never look directly at the sun, even through dark glasses. It can damage your eyes.

This is a simple sundial, like the ones you can make. Look around for sundials on buildings and in gardens.

Making a large sundial for your garden or playground

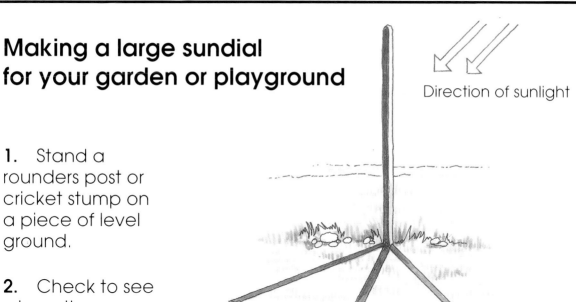

1. Stand a rounders post or cricket stump on a piece of level ground.

2. Check to see where the shadow is at different times of the day.

Making a hand-held sundial

You will need:

A piece of wood
A nail
Marker pens
A small hammer

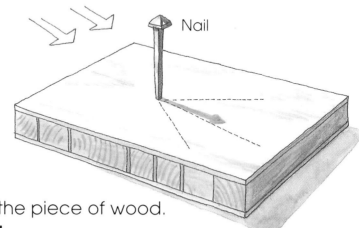

1. Hammer the nail to the piece of wood. **Ask an adult to help you.**

2. Take your sundial outside and mark on it where the shadow is at different times of the day. Make sure your sundial is always facing in the same direction.

MORE SUNDIALS

Making a cardboard sundial

You will need:

Two pieces of card
Scissors
Paper glue

1. Cut out a square piece of card. Fold it along the middle.

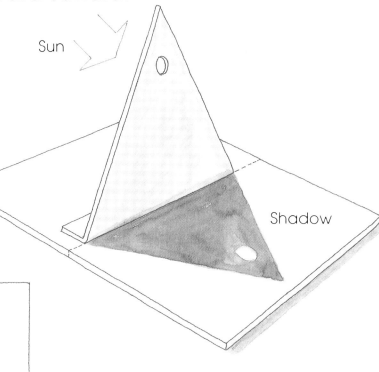

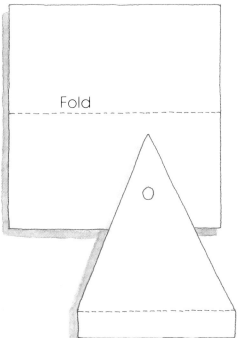

2. Cut out a triangle of card with a flap at the base. Make a small round hole at the top.

3. Glue the flap of the triangle on to the square piece of card. Fix it along the fold line.

4. Put the sundial outside and see how the shadow moves during the day.

When the sundial is not in use it can be folded up.

Further work with sundials

Place a pole in your school playing field and mark the shadow as the sun moves round. **Measure** the length of the shadow during the day. When is it at its longest? When is it at its shortest?

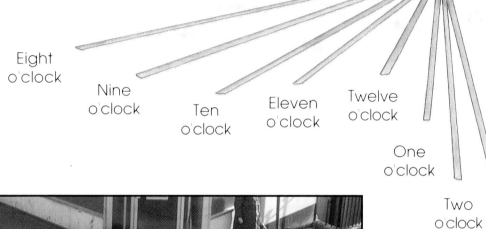

Eight o'clock
Nine o'clock
Ten o'clock
Eleven o'clock
Twelve o'clock
One o'clock
Two o'clock
Three o'clock

On a sunny day, shadows are very clear. These shadows are also very long. Can you guess what time of day this could be?

WATER CLOCKS

This is a modern water clock from the USA. The weight of the water in the side channels makes the pendulum swing.

Water clocks were used by the **Ancient Greeks** more than 2,000 years ago. The Greeks used them to keep time during the night, when they could not tell the time by the sun. The clocks were made from large bowls. The bowl was filled at sunset and the water trickled slowly out during the night.

Making a water clock

You will need:

Four plastic cups, such as small yoghurt pots
One larger plastic cup
A wooden stand
Drawing pins
A clock timer

1. Make a very small hole with a pin in the bottom of each of the smaller cups.

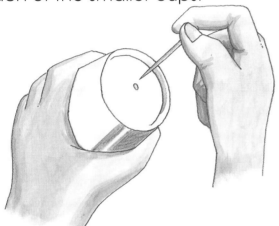

2. Use the drawing pins to fix the cups, one above the other, to the stand. Fix the larger cup, without a hole in it, at the bottom.

3. Fill the top cup with water. Measure the time it takes for the water to trickle through to the last cup.

STARTING TECHNOLOGY

OLD WATER CLOCKS

Making a Greek-style water clock

1. Make a small hole in the bottom of the yoghurt pot with the nail.

2. Put your hand over the hole and fill the pot with water. Take your hand away and let the water drip out into the bowl. Measure how long it takes to empty.

You will need:

A large yoghurt pot
A nail
A plastic washing-up bowl
A marker pen
A clock timer

3. Do the same **experiment** again. This time, mark the water level once every minute on the side of the pot to make a **scale**. Now you can use it as a kind of clock.

Chinese water clocks

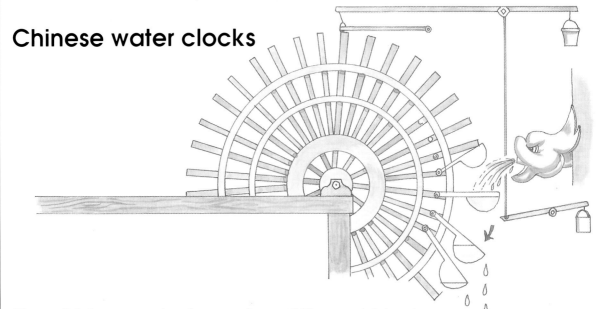

The **Chinese** designed a different kind of water clock. They used a slowly turning wheel like this. The running water turns a wheel. The wheel then turns a specially measured timer.

This very old water clock is like the ones used by the Greeks. The scale for measuring the water is on the inside.

CANDLE CLOCKS

Candles were used long ago to measure time during the night. The candles were marked into sections of one hour each. As the candle burnt down people could see how much time had passed.

The time scale on this clock is marked in Roman numbers. XII means twelve.

Making a candle clock

You will need:

A large candle
A ruler
A pin
A dish of water
A box of matches

WARNING: Never bend over a candle. Make sure your hair and clothes are away from the flame. Always ask an adult to help you make candle clocks.

1. Measure how long the candle is with the ruler.

2. Stand the candle in the dish of water.

3. Light the candle and leave it to burn.

4. After one hour, blow out the flame. When the candle is cool, measure it again to see how much shorter it is.

5. Now that you know how far the candle burns down in one hour you can measure the rest of the candle into sections of one hour each. Mark the sections with the pin.

6. Light the candle again. How **accurate** are your marks?

Stand the candle in water for safety

STARTING TECHNOLOGY

TIMERS

Sand timers have been used for hundreds of years. They are still used today. You might have one in your kitchen for measuring the time it takes to boil an egg.

Making a sand timer

You will need:

An empty plastic bottle with a plastic cap
A glass jar
Scissors
Sticky tape
Fine, dry sand or salt
A clock timer

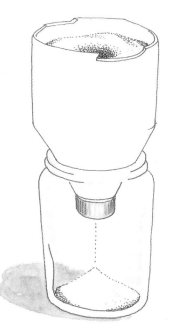

1. Make sure the bottle rests upside down in the neck of the jar.

2. Cut off the bottom of the bottle with the scissors. Punch a small hole about 3 mm across in the middle of the cap.
Ask an adult to help you with any cutting.

3. Fix the bottle upside down in the neck of the jar with sticky tape.

4. Fill the bottle with sand or salt. Measure how long it takes to empty into the jar.

Making a marble timer

You will need:

Some card
Scissors
A cardboard box
Paper glue
A small plastic cup
Marbles
A clock timer

1. Cut out three or four strips of card about 5 cm wide and 30 cm long.

2. Fold up the sides of the pieces of card so that the marbles will roll down them.

3. Glue the pieces of card on to the box like this.

4. Try to make a timer that will let you run down three marbles in exactly ten seconds. You will have to experiment with the slope of the strips of card.

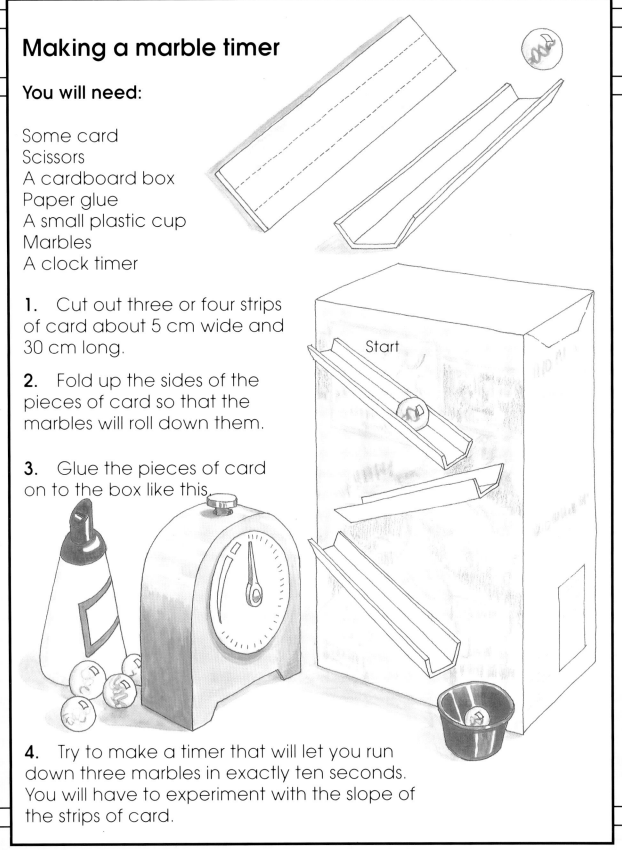

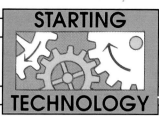

STARTING TECHNOLOGY

PENDULUMS

A ball of plasticine on the end of a piece of string makes a simple pendulum. Pendulums are used as timers in some clocks. They are very accurate timers.

You will need:

Plasticine
String
Scissors
A long stick
A clock timer

Experimenting with pendulums

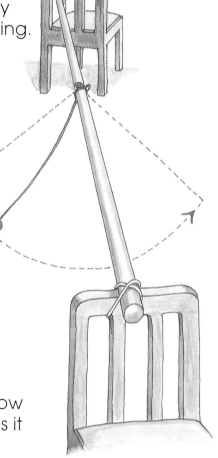

1. Make a pendulum about 1 m long by fixing a ball of plasticine to a piece of string.

2. Fix the stick between the backs of two chairs. Hang your pendulum from the stick. Make sure it can swing easily.

3. Time how many seconds your pendulum takes to do thirty swings from side to side.

4. Do the same experiment several times. Let the pendulum go from different positions. Add more plasticine to make the ball heavier. Does it always take the same time to do the same number of swings?

5. Make your pendulum shorter. Time how long it takes to do thirty swings now. Does it take less time?

Further work

A pendulum swings from side to side. Can other objects be used as timers? Try to find some. Here are some suggestions.

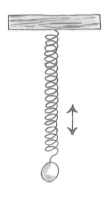

A bobbing spring

A dripping tap

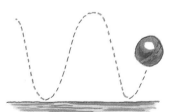

A bouncing ball

A large clock needs a long, heavy pendulum. This is Big Ben in London, England. Its pendulum is 4m long.

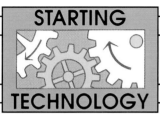

DIGITAL CLOCKS

You might have a **digital** watch of your own. It will not have hands and a clock face. It will just have numbers which may look like this: 21.00 or 08.30.

Here you can see a digital clock at a railway station. It shows the hours, minutes and seconds. When you next travel by train, see if there is a digital clock at your station.

The twenty-four hours of the day

Make a time chart like this one to show the twenty-four hours of the day.

1. With a felt tip pen and ruler, divide a long piece of paper into twenty-four sections.

2. At the top, number the sections from one to twenty-four. This is the digital twenty-four hour time.

DIGITAL TWENTY-FOUR HOUR TIME

1	2	3	4	5	6	7	8	9	10	11	12	13	14	15	16	17	18	19	20	21	22	23	24
1	2	3	4	5	6	7	8	9	10	11	12	1	2	3	4	5	6	7	8	9	10	11	12

(MIDDAY at 12; MIDNIGHT at 24)

MORNING   AFTERNOON

3. At the bottom, number the hours of the morning from one to twelve midday, and of the afternoon from one to twelve midnight.

4. Draw pictures on your chart to show what you might be doing at:
   03.00 (three o'clock in the morning)
   08.00 (eight o'clock in the morning)
   13.00 (one o'clock in the afternoon)
   19.00 (seven o'clock in the evening)

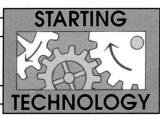

ALARM CLOCKS

We use alarm clocks to wake us up in the morning. Sometimes, people use an alarm clock when they cook food. Try to think of other times when an alarm clock could be used.

You will need:

A candle
A dish of water
A pin
String
A heavy metal object such as a metal nut
A cotton reel
A tin tray

Making a candle alarm clock

1. Stand the candle in a dish of water.

2. Attach the metal object to one end of the piece of string and the pin to the other end. Stick the pin in the side of the candle.

3. Fix up the cotton reel at the same level as the pin and hang the string over it to make a simple **pulley**.

4. Place the tray under the metal object.

5. Light the candle. When it burns down to the point where you have put the pin, the pin will come out. When the metal object falls it will make a noise!

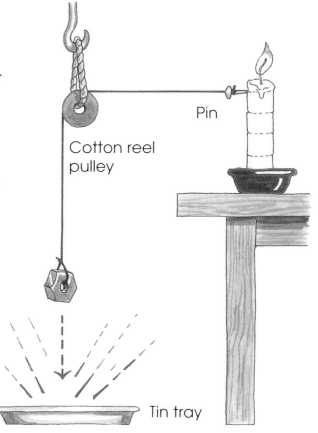

Further work

Make an electric light for your alarm clock.

WARNING: Never use mains electricity from your house. It is very dangerous.

You will need:

A piece of card
Three pieces of wire (ask an adult for insulated copper wire)
Paper fasteners
A six-volt **battery**
Light bulb and holder

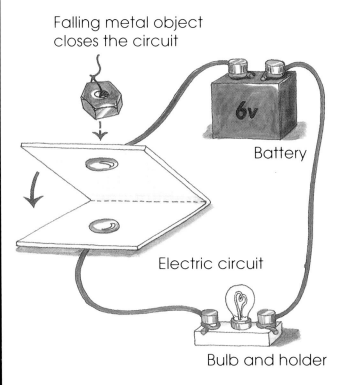

1. Fold the card in half.

2. Attach wires to each half of the folded card with paper fasteners. Make sure the fasteners only touch each other when the card is closed.

3. Connect one of the wires to the battery and the other wire to the bulb.

4. Use the third piece of wire to connect the bulb to the battery.

5. Put the card under the metal object connected to the candle. When the object falls it will close the card and light the bulb. You have made an **electric circuit**.

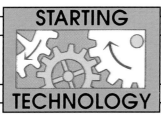

Notes for Parents and Teachers

Making the various models described in this book will involve the technological and design skills of children. However, many other subject areas can also be included. While children are actively involved in making and testing the various clocks and timers they will also be able to learn about the science, history and mathematics of time.

SCIENCE

Time is a very abstract concept. Children find it difficult to imagine time and often have little idea of age. Clocks provide the technological answer to the problems of measuring time.

Science requires very accurate clocks and timers. Teachers should be aware of the importance of allowing children to make accurate measurements of time in their science topics. For example, when testing falling objects or wheeled vehicles they will need to make a record of distance, time and speed. Even a long-term project involving simple plant growth or weather patterns will involve aspects of time measurement.

HISTORY

Opportunities exist for children to learn not only about the technological development of clocks and timers in their own country, but also about the achievements of other and ancient civilizations. They should understand the importance of time in everyday life and how people's needs may have changed throughout history. The importance of the changing seasons to farming communities is just one example of this.

MATHEMATICS

Any measurement of time will involve mathematics. However, teachers should be aware of the opportunities that such practical work allows to motivate and encourage children in this subject. Making practical estimates and calculating accurate measurements, from working models which the children themselves have constructed, may be a lot more interesting than working out sums from a book.

LANGUAGE

Language skills should develop throughout all science and technology topics. Opportunities will arise for imaginative writing as well as the more descriptive work of scientific recording. Language, spoken as well as written, should form the framework for any science topic work and will itself be enriched by the study of science. For example, the sequence of events that takes place during the actions of an alarm clock will give rise to much discussion.

National Curriculum Attainment Targets

This book is relevant to the following Attainment Targets in the National Curriculum for science:

Attainment Target 1 (Exploration of science)   The project work in this book is most important for this Attainment Target. The construction and testing of all the models in the book is relevant.

Attainment Target 9 (Earth and atmosphere)   The chapter on 'The Seasons' contains work on seasonal changes.

Attainment Target 10 (Forces)   Work on water clocks, sand timers, marble timers and pendulums involves the force of gravity. Where models have moving parts the force of friction is involved.

Attainment Target 16 (The Earth in space)   Relevant sections include work on day and night, the seasons and sundials.

The following Attainment Targets are included to a lesser extent:

Attainment Target 11 (Electricity and magnetism)   The electric circuit in the final chapter of the book is relevant.

Attainment Target 13 (Energy)   A source of energy is required for some of the models.

Teachers should also be aware of the Attainment Targets covered in other National Curriculum documents — that is, those for technology and design, mathematics, history and language.

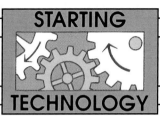

GLOSSARY

Accurate Exact and correct.

Ancient Greeks The people who lived in Greece about 2,000 years ago.

Battery A container with special chemicals in it which produce electricity.

Chinese The people who live in China. Chinese history goes back about 3,000 years.

Designs Drawings or patterns used for planning and making an object or machine.

Dial A circle with numbers round it.

Digital Using numbers instead of hands to show the time on a clock.

Electric circuit A loop of wires and objects connected up so that electricity will flow round it.

Experiment A test to find out if an idea works or not.

Measure To find out how big or heavy an object is or how long something takes to happen.

Pulley A wheel with a rope round it, usually used for lifting heavy objects.

Romans The people who ruled much of Europe and Africa 2,000 years ago.

Scale A line of regular marks used for measuring.

Season One of the four parts of the year. Spring, summer, autumn and winter are the seasons.

BOOKS TO READ

Clocks and Time by Ed Catherall (Wayland, 1982)
Time by Henry Pluckrose (Franklin Watts, 1987)
Time and the Seasons by Bobby Kalman and Susan Hughes (Crabtree, 1988)

VIDEOS AND FILMS

The Calendar: Days, Weeks, Months (Coronet)
Winnie The Pooh Discovers The Seasons (Disney Educational Productions)

Both titles are available in the UK for hire or purchase from Viewtech Audio Visual Media Ltd, Bristol.

Picture acknowledgements
The publishers would like to thank the following for allowing their photographs to be reproduced in this book: Cephas Picture Library 8; Chapel Studios 24; Eye Ubiquitous 4, 13; Michael Holford 17; Hutchison Library (J.G.Fuller) 7; PHOTRI 10; The Time Museum, Rockford, Illinois, USA 14; Tim Woodcock 23; Zefa Picture Library 18. Cover photography by Zul Mukhida.

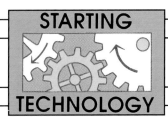

INDEX

afternoon 11, 25
alarm clocks 26-7

Big Ben 23

candle clocks 18-19

daytime 6-7
dials 5, 10
digital clocks 24-5

electric circuits 27

Greeks 14, 16, 17

hours 10, 18-19, 24, 25

marble timers 21
midday 11, 25
midnight 25
minutes 16, 24
model clocks 5
months 9
morning 4, 11, 25, 26

night-time 4, 6-7, 14, 18

pendulums 22-3
picture clocks 7

Romans 4, 10, 18

sand timers 20
seasons 8-9
seconds 21, 22, 24
shadows 10-13
sundials 10-13

time charts 25
time lines 6-7
timers 17, 20-1, 22-3
twenty-four hour time 25

USA 14

water clocks 14-17
 Chinese 17
 Greek 14, 16
weather 8-9
weeks 7

zigzag books 6, 9